Marat Khaitbaev

Innovative technologies in the automotive industry

Innovative technologies in the automotive industry

Marat Khaitbaev

Imprint

Any brand names and product names mentioned in this book are subject to trademark, brand or patent protection and are trademarks or registered trademarks of their respective holders. The use of brand names, product names, common names, trade names, product descriptions etc. even without a particular marking in this work is in no way to be construed to mean that such names may be regarded as unrestricted in respect of trademark and brand protection legislation and could thus be used by anyone.

Cover image: www.ingimage.com

This book is a translation from the original published under ISBN 978-620-7-47376-2.

Publisher:
Sciencia Scripts
is a trademark of
Dodo Books Indian Ocean Ltd. and OmniScriptum S.R.L publishing group

120 High Road, East Finchley, London, N2 9ED, United Kingdom
Str. Armeneasca 28/1, office 1, Chisinau MD-2012, Republic of Moldova, Europe
Printed at: see last page
ISBN: 978-620-7-32320-3

I dedicate this book to my favourite wife and muse - Julia."

- Marat Khaitbaev

Table of Contents

The Soviet patent engineer, inventor, writer and scientist Genrikh Altshuller was convinced of the possibility of successful experience of predecessors in sustainable and repeatable techniques of successful inventions and the possibility of teaching these methods to all modern and effective teaching methods.

To this end, a study was carried out at a cost of over USD 40,000. The study was carried out on the basis of the identified patterns of development of technical systems and invention techniques developed by the Theory of Inventive Problem Solving (**TRIZ**).

But today the problem and methods of its solution are of a principled form, because in the system of these regularities and their interconnection there are new composite materials, mobile applications and programmable controllers, control processors with elements of artificial intelligence, smart technical and technological solutions, the ability in traditional technological and technical interconnections to adapt new technical and technological interconnections, including software solutions, regardless of the fact that the purity of software solutions is not technical

When a technical problem is faced by an inventor for the first time, and especially at the start of a project, it is usually vaguely formulated and does not contain any indication of how to solve it.

In addition, as a rule, a start-up company does not have any production experience and production equipment on which to make any experiments.

In TRIZ, such a form of problems and such a form of setting the first problem is called inventive repair work. Its main disadvantage is that the engineer is faced with too many ways and methods of solution. It is labour-intensive and expensive to go through all of them, and the choice of learning methods leads to the inefficient method of trial and error.

Therefore, the first step towards invention is to reformulate the situation in such a way that the very formulation cuts off unpromising and ineffective solution paths. This raises the question, which solutions are effective and which are ineffective?

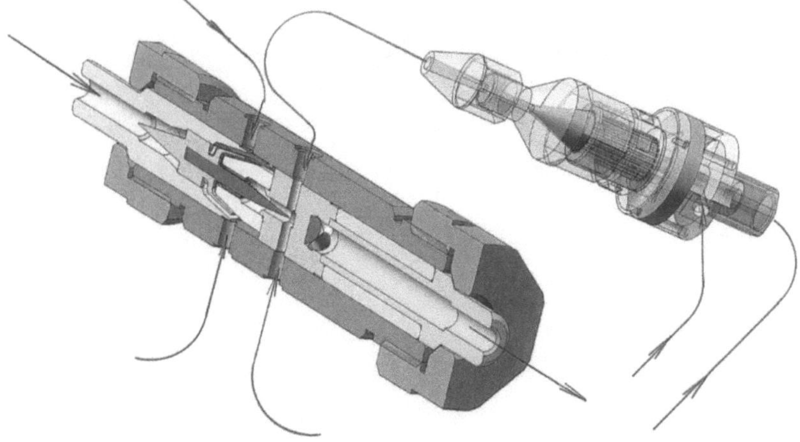

Fig.1 A variant of the solution wherein for unusual properties of the device and similar additional variants of the same device solve the problems used in the fuel system of a smart vehicle, an online continuous motion system, but this system utilises in parallel an aerosol that travels with the main fuel stream.

Г. Altshuller suggested that the most effective solution to a problem appears only at the expense of already available resources. Thus, he came to the formulation of the ideal end result (IER): "Some element (X-item) of the system or environment allows harmful effects, itself retaining the ability to carry out harmful effects".

In practice, the ideal end result is rarely fully achievable; however, it serves as a benchmark for inventive thought. The closer the solution is to the IRR, the better it is.

We propose a familiar solution close to the ideal end result (IER).

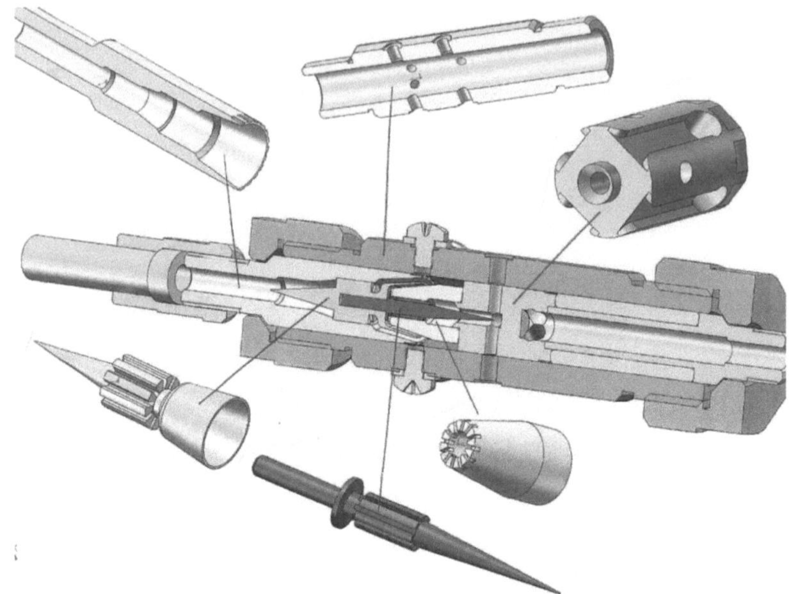

Fig.2. Device in all parameters close to the Ideal Final Result (IFR) according to the criteria of 3D design and modelling of all incoming elements in the same design and construction style.

The presented variant of design and conceptual solution allows to provide and combine the required accuracy with simplicity of layout solutions and reliability of the design with the possibility of production of all parts without exception on machine tools with numerical control when using the body of the device as a gauge, orienting all components around the central axis. In addition, this concept allows for maximum accuracy with no moving parts, again with minimal manufacturing and assembly costs.

This design option also allows the device's versatility to be protected, which in turn expands the scope of use of the device and is realised with the technology in virtually all types of ways, including those related to smart transport systems.

Having obtained a tool for cutting off ineffective solutions, we can reformulate the inventive situation into a standard mini-challenge:

"according to the ICD, everything should remain as it was, but either a harmful, unnecessary quality should disappear or a new, useful quality should appear. The main idea of the mini-problem is to consider the factors of change and the simplest solutions.

The mini-challenge formulation is accompanied by a more precise description of the task:

- What parts does the system consist of, how do they interact?

- Which connections are harmful/interfering, which are neutral, and which are beneficial?

- Which parts and relationships can be changed and which cannot?

- Which changes make the system better and which make it worse?

- How to foresee the possibility of transforming the best qualities of a future invention into a set of consumer properties, which, under favourable circumstances, can ensure the new product not only technological but also commercial success.

Let's try to model the situation sequentially.

Once the mini-task has been formulated and the system has been analysed, it is usually quickly discovered that changes attempted to improve some system parameters lead to deterioration of other parameters.

As can be seen from the examples given, the proposed design allows for multi-functional modelling and computer simulations even at the design stage, and this in principle allows for the definition of criteria for an ideal end result, and also applicable to smart transport style applications, taking into account all possible variations and conditions.

If we add to the system design with parallel computer modelling the technical characteristics of the object at different variants of structural materials and composites use, it will be possible to get pre-programmed parameters of technical characteristics of the smart transport system at the same level, for example, with or without heating of the fuel mixture flow. The figure below shows a model of such a variant of the device, providing different output characteristics with constant dimensions.

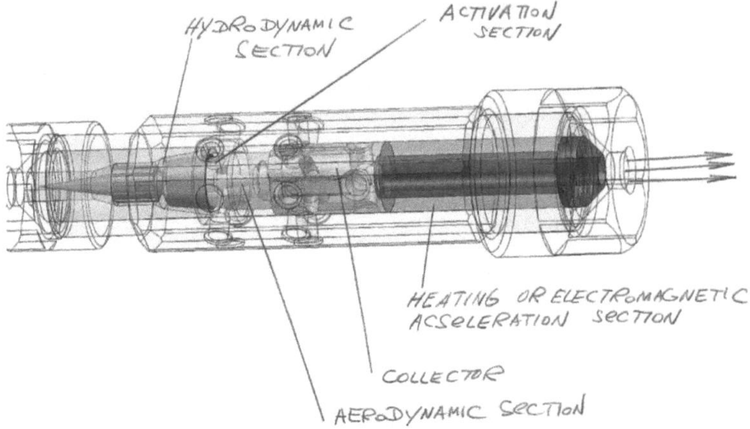

Fig.3 Simplicity of solving problems on the device, when changing the material of one of the parts, with the rest of the parameters and dimensions unchanged by a significant change, for example, the temperature of the mixture at the outlet of the device.

Contradictions of TRIZ

For comparison, it should be noted that the application of other design methods leads to contradictions. For example, increasing the strength of an aircraft wing can lead to an increase in its weight, and vice versa - lightening the wing leads to a decrease in its strength. A conflict - contradiction arises in the system.

As is known, TRIZ distinguishes 3 types of contradictions (in order of increasing complexity of resolution):

Administrative contradiction: *"we need to improve the system, but I* *don't know how (don't know how, don't have the right) to do it".*

This contradiction is the weakest and can be removed either by studying additional materials or by making/removing administrative decisions. In modern conditions, many contradictions of administrative plan are contradictions generated by commercial conditions and criteria for the realisation of the future invention, put in the technological basis of the innovative product

- **Technical contradiction**: *"improvement of one parameter of the system leads to deterioration of another parameter".*

The technical contradiction is the formulation of an **inventive problem**. The transition from administrative to technical contradiction sharply reduces the <u>dimensionality of</u> the problem, narrows the field of search for solutions and makes it possible to move from the <u>method of trial and error</u> to an <u>algorithm for solving</u> an inventive problem, which either suggests applying one or more standard technical methods or (in the case of complex problems) points to one or more physical contradictions. In the context of technical contradictions, it is necessary to see the influence on them from the commercial conditions of realisation of the innovative product

- **Physical contradiction**: *"to improve the system, some part of it must be in different physical states at the same time, which is impossible.* The physical contradiction is the most fundamental because the inventor is constrained by the physical laws of nature. To solve the problem, the inventor must use a reference book of physical effects and a table of their application.

That's the way it used to be, what's changed?

Modern computer modelling techniques, the emergence and widespread use of engineering and design software systems have given a significant addition to the system of solution techniques.

Figure 4. Details of a device in which all stages of manufacture and design are maximised to protect against the occurrence of inconsistencies while maintaining the highest possible efficiency.

System of techniques

Analyses of many thousands of inventions have revealed that for all the variety of technical contradictions, most of them are solved by 40 basic techniques.

G. S. Altshuller started to compile a list of such techniques at the early stages of the theory of inventive problem solving. To identify them, he analysed more than 40 thousand inventor's certificates and patents.

These techniques are still of great value to inventors today. Their knowledge makes it easier to find an answer in many ways, but it is no longer sufficient to solve problems on a commercial level.

These techniques only show the direction and area where strong solutions can be found. They do not provide a specific solution. This work is left to the individual.

Fig.5. The device developed taking into account the principles, techniques and methods of the Theory of Inventive Problem Solving in relation to the new requirements arising in the formation of the infrastructure of smart transport systems and their constituent elements.

With minimal dimensions, the device has a high capacity, it is able to supply 100 litres of fuel mixture per hour with a high level of homogeneity, including the level of turbulence .

The design of the device has taken into account the recommendations of simple and paired techniques to achieve an ideal end result with the necessary level of adjustment to the requirements and conditions of smart transport, smart vehicle and smart engine with smart fuel system.

The system of techniques used in TRIZ includes both simple techniques and paired techniques combined with them, also in integration with software digital technologies applied at all stages and phases of design, production and operation.

Simple techniques allow you to resolve technical contradictions. Among the simple techniques, the most popular are the 40 basic techniques discussed below.

Paired techniques consist of a simple technique and an integrative technique alternative to it and can be used to solve complex and combined physical contradictions, since two or more opposite actions, states, properties and materials, including composites, are considered.

Standards for inventive problem solving

Standards for inventive problem solving are a set of techniques that use physical or other effects to eliminate contradictions. They are a kind of complex formulas that help solve the set problems and, in turn, formulate new problems arising in the creation of new technological processes and new technological elements, including those arising on the basis of digital technologies and their widespread application.

To describe the structure of these techniques, Altshuller created a substance-field analysis.

Today, the reason for partial-material-field analysis has become computer modelling, closely and organically connected with computer-aided design methods, which in turn opens up opportunities for computer modelling of processes with elements of artificial intelligence.

A system of correspondence between classes, subclasses and related standards. This system includes 76 standards. With the help of this system it is possible not only to solve but also to identify new problems and to predict the development of technical systems.

Technological effects

A technological effect is the conversion of some technological effects into other effects. Other effects may be required - physical, nutritional, etc.

Fig.6. Real scale factor of the device with 100l/hour capacity in relation to the boat in the motor of which the specified device is installed

Physical effects

About five thousand ingredients and effects are known. Different groups of physical effects can be used in different fields of technology, but there are some commonly used ones. there are about 300-500 of them.

Chemical effects

Chemical effects are a subclass of physical effects in which only the molecular structure of matter is changed, the set of fields is limited mainly to force, velocity and heat fields. By limiting to chemical effects only, in many cases it is possible to accelerate the search for an acceptable solution.

Biological effects

Biological effects are effects changed by biological objects (animals, plants, microbes, etc.). Application of biological effects in engineering allows not only to expand the capabilities of technical systems, but also to receive and obtain results without harming nature. With the help of biological effects it is possible to carry out various operations: detection, transformation, limitation, participation of substances and fields and other operations.

Maths effects

Among the mathematical effects, the most processed are the geometric ones, which are the use of geometric shapes for various technological transformations. The use of triangles is widely known, for example, the use of a wedge or two triangles sliding towards each other.

Resources

Substantial Polish Resource (SPR) is a resource that can be utilised under the condition of problem solving or system development. The utilisation of resources increases the ideality of the system.

Laws of development of technical systems

Studying structural changes (evolution) of technical systems (TS) in time, Altshuller identified the Laws of Development of Technical Systems, knowledge of which helps engineers to predict ways of possible further improvements of products.

The laws first formulated were grouped into three combined conditional blocks:

- *Statics* - Laws 1-3, defining the conditions for the emergence and formation of TCs;

- *Kinematics* - laws 4-6, 9 determine the laws of development regardless of the influence of physical factors. They are important for the period of the beginning of growth and blossoming of TC development;

- *Dynamics* - Laws 7-8 define regularities of TC development from the influence of specific physical factors. They are important for the final stage of development and transition to a new system.

The most important law considers the "ideality" (one of the basic concepts in TRIZ) of a system.

Substantial-field complex analysis

Substance + field - a model of interaction in a minimal system, which uses a characteristic symbolism. In the conditions of smart production, smart transport, the complex analysis includes the programme factor and derivatives of this factor, taking the form - apparatus + substance + field + programme + method.

Г. S. Altshuller developed methods for analysing resources. Several of the open sources address various issues and grounds for resolving contradictions and extending the ideality of technical systems.

For example, a "teletext" system uses a television signal to transmit data by adding small time intervals between television frames in the signal.

Another technique commonly used by inventors is to analyse substances, fields and other resources that are not in use and that are observed in or near the system.

ARIZ - algorithm for solving inventive problems

Inventive Problem Solving Algorithm (ARIZ) is a step-by-step programme (sequence of actions) for identifying and resolving contradictions, i.e. solving inventive problems (about 85 steps).

ARIZ includes:

- programme;

- information supply, fed from the information fund;

- methods of managing psychological factors, which are part of the methods of developing creative imagination (CIE).

There are other approaches to help inventors unlock their creative potential. Most of these methods are heuristic. All of them were based on psychology and logic, and none of them claim to be scientific theories (unlike TRIZ):

1. <u>Trial and error</u>
2. <u>Brainstorming</u>
3. <u>Morphological analysis</u>
4. <u>Focal object method</u>
5. <u>Control question method</u>

Modern TRIZ

Modern TRIZ includes several schools that develop classical TRIZ and add new sections that are absent in the classics.

The deeply developed technical core of TRIZ (techniques, ARIZ, material field analysis) remains practically unchanged, and the activities of modern schools are mainly aimed at rethinking, restructuring and promoting TRIZ, i.e. they are more philosophical and promotional than technical.

In this regard, modern TRIZ schools often integrate and adapt digital technologies and newly emerged opportunities thanks to them at all stages and phases of project development. It is important to note that today TRIZ is actively used in advertising, business, art, early childhood development and so on, although it was originally designed primarily for technical creativity.

Classical TRIZ is a general technical version. For practical use in technology it is necessary to have many specialised versions of TRIZ, differing from each other by nomenclature and content of information funds.

Some large corporations apply TRIZ elements adapted to their fields of activity. The author believes that for application in startups, it is necessary to create a special mobile version of TRIZ adapted to the process of computer-aided design and active commercialisation.

At present, there are no specialised versions of TRIZ to stimulate discovery in the fields of science (physics, chemistry, biology, etc.).

One of the main tendencies of technical progress is the increasing struggle for author's legal developments. Therefore, there is a growing demand for personnel innovation activities and, accordingly, for methodological and software support of these activities.

Horizontal and vertical integration of all existing theories and inventions provides a dense feasibility of forming technical requirements and specifications for an autonomous power plant.

A basic layout and conceptual solution has been developed for a possible autonomous power plant, which is based on integrative technical solutions based on smart technological processes aimed at saving fossil fuels in internal combustion engines; at highly efficient conversion of energy types with simplified torque output, and the use of planar technology in low-exchange electric generator designs.

The proposed plant can be constructed according to the principle of gradual stepwise conversion of energy types, with gradual equivalent equivalent power characteristics at the output at each stage of conversion.

1. The first conversion step is the conversion of the combustion energy of the fossil fuel into mechanical power to switch on the output shaft of the internal combustion engine used as an energy converter in the first step. This provides an energy gain: utilising a fuel composite comprising a homogeneously mixed and aerodynamically suspended fuel composition of hydrocarbon fuel, ethanol and water. The proportion of ethanol and water in the recovery may be 20 -25%, in proportions of 70% ethanol and 30% synthetic water obtained from air with impurities of ethanol vapour, other alcohol or gas condensate obtained from gas engine exhaust. Deeply purified water may also be used. Methanol may be used instead of ethanol.

2. The second stage of conversion is the conversion of mechanical energy

of rotation of the output shaft of the internal combustion engine into electrical energy with certain parameters.

3. The third stage of conversion is generation of electric current pulses of certain parameters.

4. The fourth most important stage of conversion is the conversion of current pulses into torque at the output shaft of the torque amplifier converter. The amplification effect is achieved due to two factors: the first factor is kinematic, and it consists in the use of a rotor converter of linear motion into rotary motion, which allows to eliminate dead spots and energy losses associated with their overcoming; the second factor is electromagnetic, and it consists in the use for conversion of armoured electromagnets with a special design of the magnetic core and a special design of the solenoid, with additional structural advantages of the combination of the core and planar design of the coil of the soldier. All these advantages determine the efficiency of electromagnets, which belong to the category of so-called Warm or Cold magnets and have a pulling or pushing force that is an order of magnitude higher than the equivalent power consumption for its generation.

5. The fifth stage of conversion is torque supply to the shaft of the planar low-speed generator and obtaining electricity, which is the output product of the plant. The use of planar low-speed generator allows to use the advantages of power characteristics of electromagnets in their most efficient mode of operation, because the lower the frequency of operation of the electromagnet, the longer the duration of the working cycle of the electromagnet and the more time is used to restore the magnetic properties of the combined magnetic core.

All principal technical solutions underlying the proposed installation have substantial novelty and are inventions. These include:

- the use of a mixture of an organic base with an inorganic base as the combustible mixture of an internal combustion engine. Examples of such a mixture are microemulsions, nanoemulsions (if the engine has a direct injection system into the cylinders at pressures over 2000 bar; homogenised mixtures of hydrocarbon fuel with ethanol and methanol);
- application of hydrodynamic activators working on the Bernoulli principle for mixing of organic and inorganic components of the combustible mixture, carried out in a constantly moving flow of the organic component;
- the use of aerodynamic activators working according to the Bernoulli principle for foaming of composite or compound combustible mixture, carried out in a constantly moving flow of the combustible mixture;
- application for conversion of linear motion of magnet cores into rotational motion of kinematic rotor system, which does not have dead points like crank and crank mechanism in traditional kinematic systems;

- application of armoured electromagnets having a number of local technical solutions possessing essential novelty and giving the magnets significant advantages allowing to obtain an unusual effect from their operation. These are application of a combined, multilevel monolithically - twisted magnet core; application of a planar coil with a spline-shaped hole; application of a core in the form of a spline shaft; application of conductor topology on the layers of the planar coil repeating the contours of the cross-section of the spline joint.

For manufacturing and operation of the plant it is not required to develop any new technologies and new technological equipment. All units and mechanisms of the plant can be manufactured and tested on existing equipment, using known construction materials and using already used component parts.

The main components of the plant, which determine its efficiency, can be manufactured and tested separately from each other, which simplifies the manufacture and testing of the prototype, as well as significantly reduces its cost.

Figure 7. Design fragment of a new smart vehicle conceptual direction (Tesla electric vehicle), where the exterior design maximises overall convenience, allowing also for an overall combination with design elements inherent in electric and hybrid smart vehicles.

As can be seen from the presented design, the location of batteries behind the driver's cab allows, among other things, to significantly reduce aerodynamic resistance while maintaining the ergonomic parameters of the vehicle. The format of a convertible or roadster with the chosen option of battery location allowed the developers to solve the issues of optimal design of the cabin, allowing convenient and compact location of control and monitoring elements, including those with artificial intelligence functions. This design format and its basic conceptual principles may well be applied to conventional vehicles with internal combustion engines or diesel engines. The angles of the glass make it possible to create the most convenient use of the instruments for receiving and transmitting signals from navigation devices, which is also indirectly assisted by the inclined elements of other elements of the vehicle structure. This design is also optimised for the unmanned use of the smart vehicle.

Figure 8. The same Tesla car design from a different angle.

In this projection, it can be seen how the independently selected design shape provides streamlined shape constraints to all structural elements of the smart vehicle. A characteristic feature of this design shape is the absence of external corners and, in addition, special aerodynamic cavities in which the oncoming air flow is fed into special cooling channels to reduce the heating temperature of the most vulnerable components of the external and internal structure.

The dashboard of smart vehicles is extremely concise and does not distract the driver's attention, and the nature of its design allows for the need to adapt additional devices and elements on the dashboard in order to integrate them into the control and monitoring system of smart vehicles.

Figure 9. Onboard design variant of the Tesla car.

The design is an extremely streamlined system of surfaces with a convex globoid shape, which allows for the smooth formation of oncoming air flows and accompanying them into special ducts that lead the formed flow to the batteries for their intensive cooling.

Figure 10. Frontal view of the smart vehicle - Tesla electric car

The design has been developed to take maximum account of newly emerging requirements, both in terms of the positioning and efficient cooling of the batteries and in maximising the use of the intelligent control and monitoring features inherent in a smart vehicle as part of a smart transport system.

As can be seen in the figures, the design takes into account all aspects of effective formation of oncoming air flows, for which the main body of the smart vehicle is significantly wider than the cabin, which allows the oncoming air flows to be formed and channelled into special aerodynamic ducts that introduce them into the forced cooling system of the batteries, without energy consumption and without increasing drag.

In continuation of the topic of universal powertrain, it can be said that for smart vehicles, each of the variants of such vehicles - electric, hybrid, internal engine vehicles, diesel and other combined engine variants - can be adapted with smart transport chains and this is not an unsolvable task.

The illustrations of a smart vehicle shown in the figures demonstrate conclusively that, with constant consideration of all factors forming the initial technical requirements and technological factors forming the initial and regulating conditions, the present state of the art and technology allow elements and any smart devices to be incorporated into models of automotive technology.

Therefore, the following arguments can be made as a final conclusion:

- Any vehicle can be thought of as a complex energy efficient mobile installation in a system in which elements of adaptation and integration that exist today in smart conduction facilities as well as in smart events of a more moderate high organisational level are possible;

- Today's unique design concepts and rules allow the smart elements to be adapted without any changes or modifications to the construction of modern vehicles with any type and kind of engine;

-   taking into account the previous developments of local character, integrative developments of both design and construction directions are possible, and according to the availability of information as of today, there are no other restrictions on this kind of integration;

- it is really possible to use an integrative layout solution that combines all components of the system into a single energy-producing unit with full autonomy and high efficiency, which is expressed in high equivalent power generation or equivalent energy in relation to the cost of organic fuel, including biological fuel.

The main, utility-forming technical solutions of the installation have been verified experimentally, and results have been obtained to predict the effect that will result from the proposed installation when adapted into smart systems.

The new vehicle design shown in the illustrations above, which realises integrative tasks and makes it possible to come close to solving the tasks of creating a design and element base for smart vehicles in any scale factor and in any form of setting an integrative task, both within the framework of a smart vehicle only and within the framework of a complete smart transport system, including components of different technical and technological levels, united by a common innovative task of forming a complex smart vehicle.

As the statistics of project realisation show, there is an increasing impact on the commercial value and efficiency of these projects in the algorithmic component. Starting with the formulation and synthesis of innovative ideas and the finalisation of guidelines in a specific production and commercial structure.

Since the industry is currently considering experience, technologies and commercial developments in the most demanded today fuel mixture modification technologies, the author proposes to consider in various ways the algorithmicisation of this group of projects, which can also be considered an element of smart design (according to the modern classification).

At present, according to information that can be obtained from public sources, there is a tendency to modify and change internal combustion engines in directions, with changes in the automatic control systems of fuel supply and combustion. In studies related to the use of alternative fuels such as ethanol or methanol, the problem of combining ethanol, methanol and petrol or ethanol, methanol and diesel is clearly emerging.

More than a century old problem associated with the generally accepted classical system of converting linear motion into rotational motion or effect in localised aspects is not fundamental to solving the problem as a whole, and it is the mechanical problems that lead to 50% of the efficiency of any internal combustion engine .

In today's innovation community, there are filed applications for inventions that address the sets of problems described above.

The problem of supplying ethanol and petrol at petrol stations is completely solved by the invention of a prototype, manufactured and trial material, but the same prototype can be used to present, idea and solve the problem of mixing and turning mixtures into a stable emulsion, when adding fuel components such as diesel fuel and various types of industrial alcohol, including glycerin, ethanol and petrol.

In addition to all the known systems and methods of the various automatic control systems, they propose a system of non-contact control of the fuel mixture, including the levels of air saturation and the levels and nature of foaming of the fuel mixture prior to injection or on demand at the high-pressure fuel pump, as is the case in diesel engines. The creative groups proposed solutions to mechanical problems in all types of internal combustion engines, the principles of changing the mechanism for converting linear motion into rotary motion, without any change in the fuel system of the internal combustion engine.

All of the above technologies can be widely applied if a number of fundamental conditions are fulfilled.

To ensure a timely and complete combustion cycle in a short period of time, the fuel must fulfil the following requirements:

1) good permeability in fuel pipework to ensure reliable operation of the high-pressure fuel pump (at unstable viscosity of 2-6 mm2/s and temperature of 20 °C) and absence of mechanical impurities and water;

2) necessary atomisation, good mixing and vaporisation, for this purpose the fuel must have an optimum viscosity and a certain fractional

composition;

3) the necessary ignitability for easy starting of a cold engine, smooth pressure build-up and complete smokeless combustion (these properties are related to the chemical and fractional composition of the fuel, as well as viscosity; the chemical composition of the fuel is measured by standard numbers, which characterises ignitability and is the main feature of the fuel's engine properties);

4) permissible level of soot and other deposits formation on valves, rings, pistons, prevention of atomiser needle coking (propensity of fuel to soot depends on chemical and fractional composition, viscosity, content of mechanical impurities and water);

5) absence of corrosive products (corrosive properties of fuel depend on the presence of mineral and organic acids, sulphur compounds and water);

6) higher calorific value;

7) high fuel flowability with minimum leakage through the gaps in the plunger pairs with the effect of wear of rubbing vapours.

When soft-air diesel, without water, is switched on, corrosion processes occur as follows:
The first version is the enhanced addition of diesel fuel with water and the subsequent increase in the amount of blends of diesel fuel with water and compressed air. After rotation with compressed air, the mixture contains air bubbles uniformly distributed throughout its volume, with a diameter of 20

micrometres, with a constant pressure of 20 bar and each of the bubbles has a 20 micrometre thick shell consisting of an emulsion containing 92% diesel and 8% water. After injection into the combustion chamber, the mixture is atomised into particles of 3-5 micrometres in size and the diesel and oxidant form an environment in which the organic combustible element is distributed between the oxidant, allowing ignition at lower temperatures and creating a quality equivalent number of 45 to 50.

The second version of the technology represents the use of only diesel fuel with compressed air and, while maintaining the conditions of the first version of the technology, the ignition of the fuel mixture and the qualitative equivalent numbers maintaining qualities similar to the first version.

Technological instruction

The process of manufacturing original parts and selection of standard parts, materials and components in the manufacture of prototypes of a complex device for activating fuel mixtures in groups of internal engines used as fuel petrol and diesel engines.

1. Technical requirements for the manufacture of original parts

1.1. All original parts must be manufactured in full conformity with the working drawings

1.2. All tapered surfaces of the parts shall be experimentally tested on aluminium mock-ups, with the mock-ups of the mating parts used as reference templates and gauges

1.3. Finishing of parts after machining in the form of electrochemical polishing, should be carried out only after checking all parts included in the device for assemblability

1.4. Heat treatment of parts must be carried out in the workpiece, in accordance with the requirements of the technical documentation

2. Technical requirements for the assembly of the device

2.1. All sections of the unit must fit into the opening of the outer casing in a sliding fit

2.2. The manifold casing, which has 8 independent channels, 4 of which are for compressed air supply, must be aligned during assembly with the holes in the outer casing, through which compressed air is supplied from the compressor.

2.3. After alignment, screw the clips into the holes of the outer casing and then tighten the nut of the outer casing; after tightening, unscrew the clips one by one and install the fittings fixing the air pipe one by one.

2.4. The order of assembly is as follows:

- A nipple is placed on the fuel line;

- The nipple is attached to the pipeline;

-   The resulting connection is connected to the nut of the outer casing and put on the body of the hydraulic section together with it;

- The entire resulting connection is inserted into the bore of the outer casing and clamped with a nut;

- After that, the conical interface of the hydraulic section is inserted from the opposite open side of the outer casing, after which the interface of the pneumatic section is pressed into its centre hole, the body of the pneumatic section is slid onto its slotted part and the seat of the manifold body is inserted therein

2.5. The collector casing is orientated according to point 2.3;

3.   Technical requirements for standard fuel pipework elements

3.1. The standard device includes fittings for attaching the elements of the 4 pipes supplying the extended air from the compressor, 4 metal tubes connected according to the dimensions of the fittings; two tubes for cutting the fuel line before and after the device, against which the dimensions of the two nipples must be calibrated and to which these nipples must be hermetically attached

4. Technical requirements for compressed air pipework from the compressor

4.1. The pipework for supplying compressed air from the compressor to the fuel mixture activator, complies with the requirements of 3.1.

5. Technical requirements for the compressor

5.1. For the initial test phases, the compressor type is not of fundamental importance

5.2. The compressor must have a rated operating pressure of at least 20 atmospheres

5.3. An air filter with a liquid buffer must be installed at the compressor air inlet

5.4. The compressor must have at least 4 compressed air outlets, or a system for uniformly dividing the compressed air flow at the compressor outlet

5.5. The pressure level must be adjustable over the entire technical range of the compressor from minimum to maximum

5.6. The compressed air flow rate at the compressor outlet must be controlled from maximum to minimum

5.7. All outlet pipework connections to the compressor must be standardised

6. Programme and methodology of preliminary tests at the device manufacturer

6.1. Each part must have a quality certificate

6.2. Each part must be accompanied by a record of the conformity of the linear dimensions of the part with the dimensions on the drawing.

6.3. All parts must be selectively chosen for each assembly

6.4. For each assembly, a test report shall be drawn up to verify that the assembly is assembled and meets the general technical requirements of the product

6.5. All connections of the hydraulic lines of the unit, both at the inlet and outlet of the unit, must be leak tested at a pressure exceeding the nominal operating pressure by 25%

6.6. All connections of pneumatic lines of the device should be checked for tightness under the pressure in the pneumatic system exceeding the nominal pressure by 25%

6.7. Fully assembled device should be checked for tightness of both hydraulic and pneumatic sections, when they are simultaneously operated at a pressure exceeding the nominal working pressure by 25%

6.8. A manufacturer's data sheet shall be prepared for each device showing the results of all preliminary tests

7. Technical requirements for an air filter to be installed at the compressor inlet to which the fuel mixture activation device is connected

7.1. The compressor must have an operating pressure of at least 20 atmospheres

7.2. The pressure must be adjustable between minimum and maximum values

7.3. The air cleaner must operate at the specified parameters

7.4. The filter must have a system for conducting airflow through a liquid such as light mineral oil or industrial alcohol

7.5. The air inlet to the filter must be via tangential inlet ducts

Considering the rather compact version of the proposed device and its easy incorporation into the fuel system of a modern engine, including that built into the engine of smart vehicles, permanently embedded in complex smart transport systems.

The author believes that due to the exceptional universality of the idea, the condition is the necessity to classify the formulated idea and its constructive embodiment as soon as possible, because in the market of intellectual products it will enter simultaneously with the basic product, including for means of smart transport, - various variants of biological fuels. mixtures and various operations of homogenisation, including liquid food products.

A characterisation of the field of claim which can be declared today: A mixer embedded in a pipeline, which utilises for mixing the pressure differences occurring or created in different sections of the pipeline due to differences in linear velocities of the flow of fluids in said pipeline;

Hydrodynamic mixer;

Hydrodynamic mixer integrated with foam generator;

Mixing and, at the same time, foaming unit with sequential action;

A sequential, in-line, mixing and at the same time foaming unit that utilises the kinetic energy of the flow of fluids moving in said pipeline to perform its technological functions;

Hydrodynamic turbulence stabiliser-mixer;

Hydrodynamic turbulence stabiliser-mixer integrated with aerodynamic turbulence stabiliser-foam generator;

Hydrodynamic turbulence stabiliser for fluids moving in a pipeline;

In-line, sequential hydrodynamic and aerodynamic stabilisers of turbulence in the flow of fluids moving in said pipeline;

Aerodynamic foam generator;

Compact, aerodynamic foam generator integrated in the fuel line;

Hydrodynamic mixer of different viscosity and chemical-physical properties of liquids;

A hydrodynamic mixer of liquids of different viscosity and chemical-physical properties, integrated into the fuel pipeline, compact, not requiring additional energy costs for functional operations;

Hydrodynamically and aerodynamically integrated, compact, in-line unit for the preparation of fuel mixtures before their injection into the combustion chambers of combined engines realising the Otto cycle.

Determination of the main distinguishing characteristics of the complex device for activation of fuel mixtures as part of the basic component of smart technology and part of the complex of smart technologies included in the complex - smart transport and smart transportation

1. The device has a complex combined effect on the fuel mixture and its constituent components and does not require any, even minimal, modifications to the fuel system of a smart vehicle for incorporation into the fuel system of the vehicle

2. The device has several successive functions of the type of conversion and properties of the fuel mixture flow, wherein all said conversion steps drive the device into a state of constant movement of the starting materials and additional components of the fuel mixture

3. The device in the process of influencing the components of the fuel mixture, has the ability to work simultaneously with liquid and gaseous media, that is, at any moment of the working cycle of the device there is a simultaneous synchronous synchronous impact on the liquid and gaseous component of the fuel mixture, and each of the components in turn erects and affects the characteristics of other components and the final characteristics of the fuel mixture when it is injected into the combustion chamber, which is an introduction of smart technologies within the framework of smart vehicle.

4. The influence of components in the composition of the activated fuel mixture on its properties and characteristics, on the conditions of use of these fuel mixture components in the smart transport system and its efficiency are preserved and after injection into the combustion chamber of the smart vehicle as part of the smart transport system

5. The introduction of both additional and gaseous components into the fuel mixture flow is carried out through tangential channels with the formation of a vortex effect and further utilising the identified vortex effect towards the vortex tube or equivalents thereof.

6. The complex activation device includes 10 successive, additional steps of transforming the shape and cross-section of the fuel mixture flow, which are only for the purpose of achieving the intended purpose and attempting to tune each of them individually, without functional relationship to the coronaviruses, which are erroneous.

7. The purpose of the device is not to increase the level of turbulence, but a complex effect, which includes several key technological techniques, such as transforming the shape and characteristics of the flow at the inlet, creating a zone with a high level of local turbulence and the introduction of additional liquid fuel components in this zone of the vortex effect path. , then carried out in the same zone of the second rarefaction zone, by introducing a compressed gas flow with a pre-formed turbulence level and vortex effect, which completes the activated fuel mixture and ensures its dispersion determination after injection into the combustion chamber; All the opposite technological transitions basically build a comprehensive smart process within a comprehensive smart vehicle within a smart transport system.

8. It is known from the laws of physics that in a pipeline the movement of liquid in contact with the pipeline walls has a developed turbulent structure. For the fuel pipeline this situation is aggravated by the fact that in most cases at the end of the fuel line there is a high-pressure pump, developing at the outlet pressure of 2000 bar and more, and the movement of fuel has some cyclic character. This factor turns the process of fuel homogenisation by turbulence level into a guarantee of fuel system stability and ensures the possibility of its normal and optimal application in smart transport systems.

9. In the device, the process of transforming the shape of the flow cross-section has the purpose of transforming the shape of the flow from circular to circular, which allow 2.5 times increase the contact perimeter, and, accordingly, increase the turbulent characteristics of the flow in those sectors of the cross-section of the flow (especially in the centre of the cross-section of the flow), in which under normal condition the level of turbulence - the minimum.

10. Since the compressed gas is introduced into the device in a state of developed turbulent flow and according to the vortex principle, the level of accumulated turbulence and kinetic energy in the fuel mixture increases exponentially. Certainly, when applied in smart vehicles, this factor should be taken into account in problem statement and preparation of technical requirements and relevant specifications

11. Since the compressed gas flow is introduced into a hermetically sealed volume at an initial pressure of (approximately) 20 atmospheres, and at the inlet to the device forms a local annular zone having the main features of the local rarefaction effect corresponding to Bernoulli's principle, air bubbles are detached from the flow and covered with a shell of liquid components of the fuel mixture, and the internal pressure in the bubbles, due to the fact that the liquid is not compressed, increases due to the fact that during the pause

between injections air continues to flow, the number of bubbles increases. This phenomenon plays an extremely important role for applications in smart transport systems as a factor in ensuring a sufficient level of environmental safety.

General analysis of the current state of the art in fuel mixture activation technology for internal combustion engines installed in smart vehicles

Various technologies and devices are now known for the ultimate goal of reducing fuel consumption in vehicle engines.

According to the nature of the effect on the fuel, it is possible to distinguish a number of technologies in which the activation of the fuel and the resulting increase in its efficiency is carried out by means of various magnetic devices.

These are inventions under US patents. In all of them, a magnetic field is used to activate the fuel. This action is mainly performed when the fuel is fed into the combustion chamber at one or more points, or localised sections of the fuel line. Such an effect is not direct, which determines the relatively low efficiency of such methods.

There are also known examples of effects not only on the fuel, but also on the air mixture in combination with fuel effects. All these variants have a local character and retain their influence on the activity of the fuel or fuel mixture for a very short period of time. In addition, it is problematic to maintain any influence of magnetic activation methods directly on the combustion process, which ultimately has the greatest impact on the fuel efficiency and combustion completeness. The completeness of combustion is also the main criterion determining the level and concentration of toxic substances in the exhaust gases.

Other methods for reducing fuel consumption are also known: ensuring a constant percentage of petrol to air ratio; creating fuel flow damping systems; and using a dual fuel heat exchanger. Numerous modifications to fuel injectors and injectors are also known, the result of which is a slight improvement and increase in the efficiency of the internal combustion engine.

All of the above methods of influence have one similar process detail - they all have no direct effect on the properties of the fuel or fuel mixture and their impact and influence on the combustion process itself is very small or non-existent.

Commercially available devices for increasing fuel efficiency and reducing specific fuel consumption are also known. These are groups of products and devices called: FUEL ENERGIZER, FUEL VAPORIZER and TORNADO FUEL SAVER.

It makes sense to dwell on the latter in more detail, since this method and in general the technology of vortex influence on the components of fuel or fuel mixtures have a limited, but still direct influence on the combustion process. The effect on the fuel mixture or on the individual components of this mixture is carried out in a known device by swirling the air flow, which is mixed with the fuel flow, in most cases petrol. Swirling increases the degree of turbulence in the fuel flow, and this in turn increases the level of combustibility, i.e. due to this more energy is obtained from the burnt volume of fuel with more complete combustion, which determines the reduction of specific fuel consumption.

In the proposed technology of special treatment of the combustible mixture during its movement in the fuel pipeline, several consecutive methods of influence on the moving flow of fuel and then fuel mixture are used, using the basic physical principles of hydrodynamics and further aerodynamics Completion of the process of the proposed impact is carried out by means of multiflare nozzles, which do not change the physical and chemical properties of the fuel mixture when directly fed into the combustion chamber.

The proposed technology has the possibility of hydrodynamic mixing of organic and inorganic components of fuel with subsequent homogeneous aerodynamic oxygen saturation, up to full or partial foaming, or highly effective homogenisation of turbulence turbulence of homogeneous fuel flow with also subsequent homogeneous aerodynamic oxygen saturation.

In both the first and second embodiments, when injected directly into the combustion chamber, all physical properties of the fuel and chemical properties of the fuel resulting from the exposure are retained.

The proposed technology and two embodiments of the device for its realisation are as follows:

1. Technology

Fuel systems of vehicles with conventional engines and energy systems of electric vehicles of all types.

The proposed technology of influence on the dynamic mobile fuel flow in the fuel pipeline consists of several stages of deep hydrodynamic and aerodynamic influence on the fuel. In this case, to increase the depth of impact and to reduce or completely exclude additional energy costs for impact, conditions are created for the operation of the Bernoulli principle in the pipeline, which allows at minimum energy costs at the final stage of aerodynamic impact and absolutely without any costs at the stage of hydrodynamic impact to obtain a highly efficient fuel mixture and significantly increase the efficiency of the combustion process and at the same time reduce the specific fuel consumption for it.

2. Device options

The devices for implementing the proposed technology have two designs: First, for cases where the fuel has at least two heterogeneous components (e.g. petrol and water) or where the fuel has at least two homogeneous components (e.g. petrol and ethanol) or where the fuel has at least three heterogeneous components (e.g. petrol, ethanol and water).

The second version is intended for cases where the fuel has only one component - e.g. petrol.

For all versions, the device consists of a hydrodynamic turbulence homogenisation mechanism integrated in the pipeline connected functionally and structurally by means of a hydromechanical and aeromechanical interface, associated with a fuel supply pump from the fuel tank and an aerodynamic foaming flow activator associated with a mini compressor having a rotational drive from the kinematic branch of the engine shaft and through the fuel pipeline having an outlet to the multiflare fuel injectors.

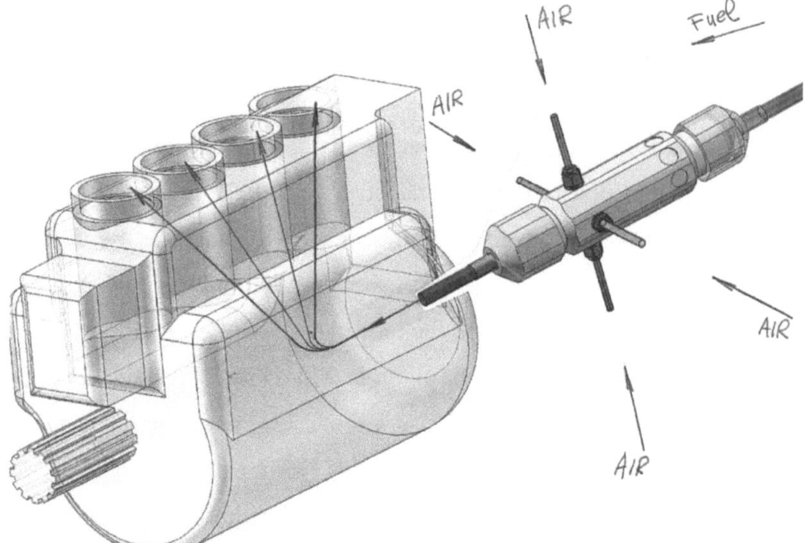

Figure 11. Three-dimensional model of the engine with fuel-activating device

Fuel activation is performed by homogeneous mixing of liquid fuel with microscopic air bubbles. The mixing is performed in the pipeline during the movement of liquid fuel from the low-pressure fuel pump to the high-pressure fuel pump. In this case, air is supplied from four sides to the area of the device in which a low-pressure zone is formed during the movement of fuel through the coaxial channels of the device, and in some cases a vacuum is formed in this zone. In this mixing embodiment, the size of the air bubbles does not exceed a few micrometres, and the thickness of the outer spherical shell on each bubble also does not exceed a few micrometres. After injection, the air bubbles rupture, increasing the dispersion of fuel particles in the engine combustion chamber.

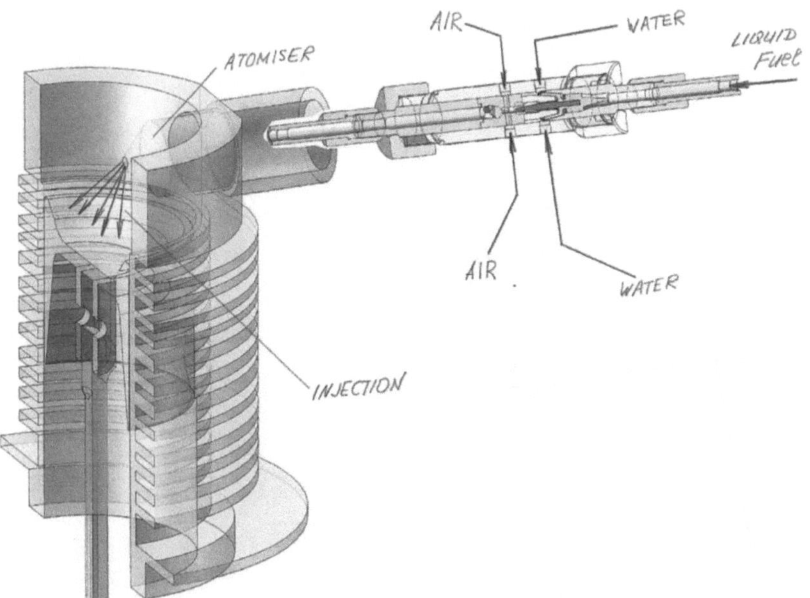

Figure 12. Three-dimensional model of an engine cylinder with the activating device inserted directly into the cylinder cavity.

This variant of fuel mixture activation provides in the device the first operation - mixing of fuel with water and the subsequent operation - mixing of the obtained fuel emulsion with compressed air according to the method and technology shown in Figure 12.

It is important to note the fact that in this method of activation the fuel mixture dramatically changes its properties in at least two stages of activation. The first stage is the online preparation of the fuel emulsion, in which the emulsion is an incompressible liquid (typical liquid properties characteristic of fuel and the same properties characteristic of an emulsion - a mixture of two liquids). The second step is online gasification of the emulsion, in which the gasified liquid - the emulsion - is a compressible liquid. At injection of such compressible liquid - emulsion - dispersity of atomisation has minimal parameters and combustion process is the most efficient.

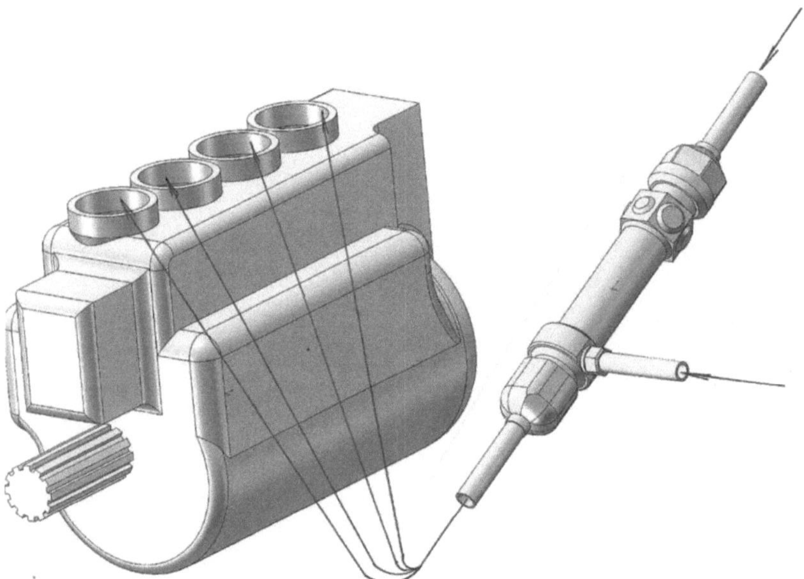

Figure 13. Three-dimensional model of an internal combustion engine with an activating device in which the activating element is introduced through a tangential nozzle, which is introduced into the mixing zone in the area of coaxial ring channels and local rarefaction or vacuum.

The same activation device can be used for online homogenisation of the fuel flow. If we consider this activation configuration as a means of ensuring the reliability of the engine fuel system, the simplicity of this system, its reliability, small size and other tangible advantages allow it to be adapted to virtually any configuration of a smart vehicle system without any restrictions and modifications to the engines and fuel lines of smart vehicles.

For these configurations, the need for only one activation device per engine makes it much easier to incorporate smart transport into the transport system of a metropolitan area.

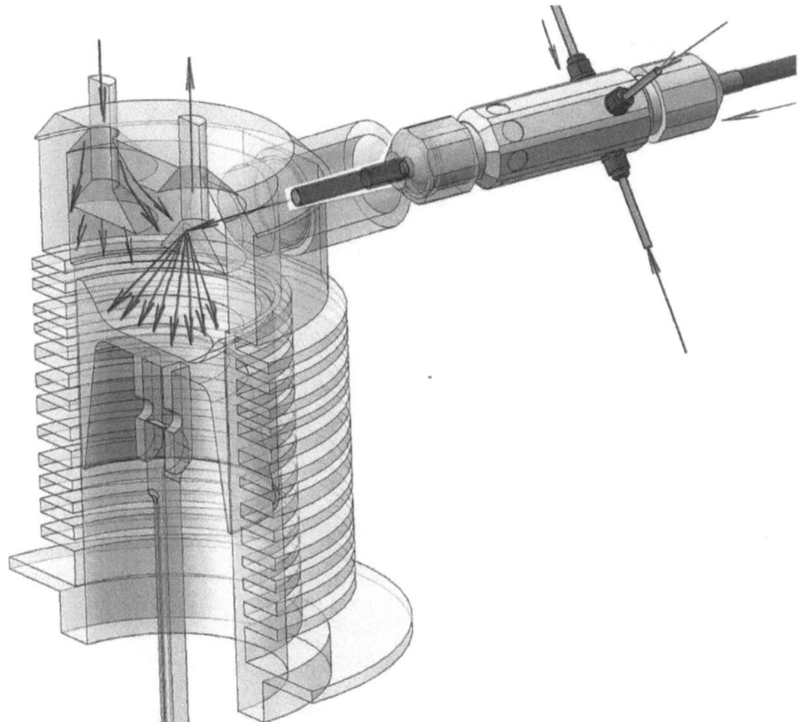

Figure 14. Engine cylinder into which the outlet pipe of the activating device is introduced, which may have several modifications: homogenising element or online emulsion preparation device or online fuel blend preparation device .

In all variants, air in stoichiometric proportions is introduced directly into the combustion chamber and the exhaust gas streams are discharged from a cavity at the top of the cylinder. This variant is closest in principle and design to conventional internal combustion engine cylinders and diesel engines.

43

Let's return to the Theory of Inventive Problem Solving (TRIZ) and consider **40 known principles in the** context of using techniques and methods inherent and characteristic of smart manufacturing, smart transport and smart design.

1. Crushing principle

- to divide the object into independent parts;

- to make the object collapsible;

- to increase the degree of crushing of the object.

The criterion of independence of parts into which it is recommended to divide an object in modern machines and apparatuses is practically impossible to fulfil. If we take into account the fact that modern innovative objects most often represent an integrative combination of apparatus, system, programme and method, it becomes clear that all parts or components of the object are to some extent tied to these elements. If we follow this logic, then it turns out that if it is necessary to achieve complete autonomy and independence of the parts of the object, it is necessary to endow each part with the correspondence specified to the apparatus, system, programme and method, which, given the requirement of the patent law on the indivisibility of the object of invention and knowing the design principle of the meaninglessness of repeating all the design, software, technological and algorithmic features identified in the process of dividing the object into parts in each of the parts, makes this technique not so useful

The methods of modern innovative design are full-scale integration, and the principle of fragmentation should most likely turn into the principle of selecting autonomous functional components of an object for horizontal and vertical layout integration.

Since pure invention does not exist in the modern innovation process, at least one of the elements of integration should be the principle of commercial feasibility for horizontal integration and the principle of implementation universality in various technological categories, not always at first glance logically related to each other, for vertical integration.

The Theory and Algorithm of Inventive Problem Solving was created at a time and in a situation when there was no production of universal standardised design and technological components in the economic system. This determines the fundamental difference between modern machine design and design and the design methods and criteria that existed at the time of the Theory and Algorithm of Inventive Problem Solving.

First of all, it is necessary to define the new principles of machine design inherent in intelligent technical solutions in the field of transport and vehicles.

Today, the essential building blocks of this system of interconnected technical solutions and associated software products and mobile applications represent an integrative concept for the application of software products in the technological cycle of basic technological entities, especially in vehicles, such as engines of all types, fuel technology, environmental systems, real-time control and monitoring of their combinations.

2. Principle of judgement

- to separate the extra part from the object;
- to pick out the only part you need.

In smart transport systems, such rendering is the functions of computer modelling in real time, including at the expense of a mobile application integrated into the specified transport system. The results of such modelling make it possible to significantly approximate the modes of operation of the transport system and each vehicle in its composition to the systems containing elements of artificial intelligence.

3. Local quality principle

• to move from a homogeneous structure of the object (or external environment, external influence) to a heterogeneous one;

• different parts of the object should have (fulfil) different functions;

• each part of the facility should be in the conditions most favourable to its operation.

In such a system, the most favourable operating mode of the facility is determined by computer simulation results of the optimal model of the smart transport system and can be transformed to each vehicle within that system.

4. Principle of asymmetry

• to go from a symmetrical shape of the object to an asymmetrical one;

• if the object is asymmetric, increase the degree of asymmetry.

This principle preserves the autonomy of each vehicle within a smart transport system and ensures the highest possible level of integration of each vehicle into the overall objectives and characteristics of the smart transport system

5. Unification principle

• to connect objects that are homogeneous or intended for related operations;

• combine homogeneous or related operations over time.

In a smart transport system, all the constituent elements are smart vehicles which, each individually and all together, under the control, monitoring or adjustment carried out by the mobile application, can combine vehicles performing homogeneous operations or separate vehicles or groups of vehicles performing heterogeneous operations, but the qualification of the degree of homogeneity or heterogeneity is performed by means of the mobile application.

6. Principle of universality

- the object fulfils several different functions, thus eliminating the need for other objects.

In this situation, the question of the appropriateness of an object to fulfil functions inherent in other objects becomes obvious and legitimate .

Who needs it, who would go to the trouble of combining the functions of other tools in one product, such as a tool. Such a tool can be used by only one operator anyway, and if at the same time it is necessary to use a new function included in the invention, it is necessary to have one more such tool, i.e. commercially such an invention is of no use and will not be supported by investors.

This principle is quite different in intelligent complex systems, where the unifying function is primarily implemented by a mobile application that solves complex monitoring, control and measurement tasks and at the same time contains all the necessary interchangeable information for real-time computer simulations.

7. The Matryoshka doll principle .

- one object is placed inside another, which in turn is inside a third, and so on;
- one object passes through cavities in another object.

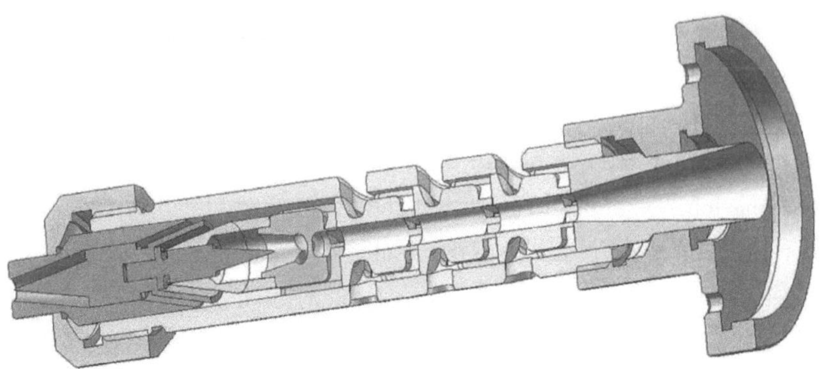

Fig.15. Real design of fuel mixture activation device in carburettor engines based on the matryoshka doll principle.

The design principles, similar to the general design principles of devices for activating fuel mixtures, are characterised by the use of the outer body of the device as a gauge defining the position of each internal element.

All internal cavities and channels of one component are connected with similar channels and cavities of adjacent components, and constructive elements of vortex generators are connected geometrically and functionally with suction windows of the outer casing, i.e. all features of a coaxial matryoshka doll are present, in this particular case determining the quality and efficiency of the whole device in relation to the circumstances of the system of a higher level, in turn ensuring compliance with the requirements for smart transport systems.

8. The antivenom principle

• compensate for the weight of an object by coupling it with another object that has a lifting force;

• compensate the weight of the object by interaction with the medium (due to aerodynamic and hydrodynamic forces).

In this case, in the current environment, options are available related to the potential use of different composite materials with different specific gravities.

Today's methods and machine design programs allow real-time simulation of the weight of any structural component with analysis of the reaction of adjacent structural elements to the weight characteristics of interacting parts and their mutual influence on adjacent assemblies and joints.

9. The principle of pre-emptive antidote

• give the object stresses opposite to unacceptable or undesirable operating stresses in advance;

• If by the conditions of the task it is necessary to perform some action, it is necessary to perform the counteraction beforehand.

At the initial definition and classification of this principle, there were no computer modelling capabilities, which made it very common to perform many actions and operations during design that today are performed automatically during design using real standard computer engineering software.

With the help of these programs, actions and technical variants are optimised during the design process and the most optimal variant is entered into the project, which takes into account all circumstances and initial technical requirements.

10. Pre-action principle

• perform the required action (fully or at least partially) in advance;

• arrange the objects in advance so that they can come into action without time-consuming delivery and from the most convenient location.

As there are no moving parts in the activation device, the operating cycle of the activation device is precisely the execution of an action without time lost in positioning the constituent parts of the device.

11. The "pre-padded cushion" principle

• compensate for the relatively low reliability of the facility with pre-prepared emergency means.

From the point of view of modern relations between investors and inventors, it is clear that a low reliability of an object will safely and firmly close such an object to the market. Only a tested and absolutely workable design determines the real commercial success of an object.

12. Equipotentiality principle

• change the working conditions so that you do not have to lift or lower the object.

The absence of moving parts in the construction of the device confirms full compliance with this principle.

13. The "backwards" principle

• Instead of the action dictated by the task conditions, perform the opposite action;

• to make the moving part of an object or environment stationary and the stationary part moving;

• turn the object upside down, twist it.

This is the principle realised in the device for activating fuel mixtures - activation takes place without the use of moving parts.

14. The principle of spheroidality

• move from rectilinear parts to curvilinear parts from flat surfaces to spherical surfaces, from cube and parallelepiped parts to spherical structures;

• use rollers, balls, spirals;

• to change from rectilinear to rotational motion, to use centrifugal force.

As an equivalent to the principle - using only cylindrical shaped parts.

15. The principle of dynamism

- the characteristics of the object (or external environment) must change so as to be optimal at each stage of operation;
- to divide an object into parts that can move relative to each other;
- if an object is generally stationary, make it mobile, moving.

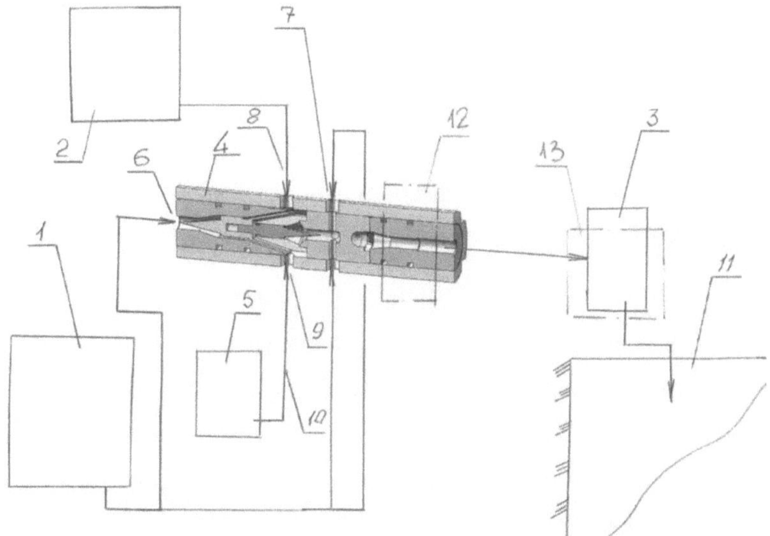

Fig.16. Combined system for preparation of innovative fuel blends at fuelling stations with elements of dynamic influence on fuel blend components as an illustration of variants of realisation of the principle of dynamism in complex infrastructure systems of smart transport systems.

The numbers in the figure denote:

1 - base fuel tank

2 - tank with auxiliary fuel or fuel mixture component

3 - storage tank with homogenisation systems

4 - fuel composition preparation device

5 - tank with additional fuel mixture component or water for emulsion preparation

6   - fuel input to the device

7   - fuel inputs to the activating device during homogenisation

8   - fuel component input

9   - fuel component input

10 - fuel component supply line to the device

11 - petrol station fuel tank

12 - control area

13 - control area

16. Principle of partial or redundant action

• If it is difficult to get 100% of the required effect, it is necessary to get "a little less" or "a little more" - the task will be considerably simplified.

In fact, for a new innovative product that implements a new process or principle, determining the magnitude of the real effect is one of the most important analytical problems that can practically be solved by numerous computer models and their modifications that take into account all aspects of the new development. Computer modelling in this case should take into account all the results of preliminary and control tests of the innovative product and should take into account the control results of qualification tests of similar products.

17. The Principle of Transition to Another Dimension

• difficulties related to the movement (or placement) of an object along a line are eliminated if the object acquires the ability to move in two dimensions (i.e., in a plane). Correspondingly, the problems associated with the movement (or placement) of objects in one plane are eliminated when moving to space in three dimensions;

• use a multi-storey layout of facilities instead of a single-storey layout;

• tilt the object or put it "on its side";

- to use the reverse side of this square;

- use optical flows falling on a neighbouring area or the reverse side of an existing area.

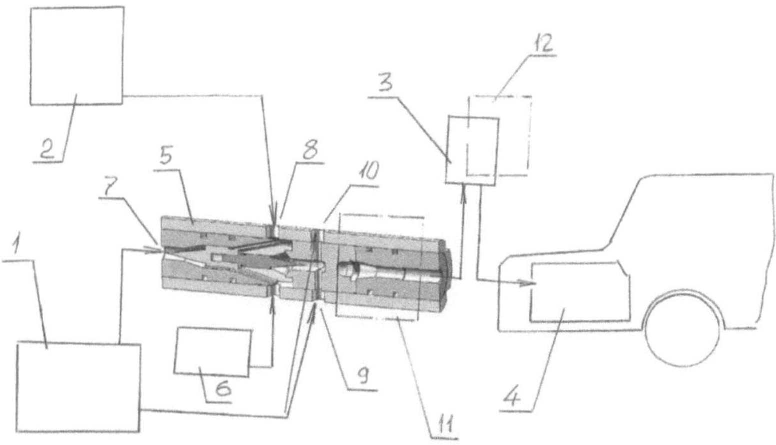

Figure 17. Combined system for preparation at fuelling stations and delivery to the tank of a car or other vehicle of fuel mixtures, emulsions or homogenisation of fuel before delivery to the vehicle tank

The numbers in the figure denote:

1 - base fuel tank

2 - fuel component tank

3 - tank - a storage tank before fuel or fuel mixture is fed into the tank of a car or other vehicle

4 - the fuel tank of a car or other vehicle

5 - device for activating fuel mixtures or on-line homogenisation of fuel mixtures or preparation of fuel emulsion

6 - auxiliary fuel mixture tank

7 - fuel input to the device

8 - introduction of fuel mixture components into the device

9 - introduction of fuel mixture components into the device

10 - introduction of fuel mixture components into the device

11 - control area

12 - control area

18. Use of mechanical vibrations

- to set the object in oscillating motion;

- If such a movement is already taking place, increase its frequency (up to ultrasonic);

- use the resonant frequency;

- to use piezoelectric vibrators instead of mechanical vibrators;

- use ultrasonic vibrations in combination with electromagnetic fields.

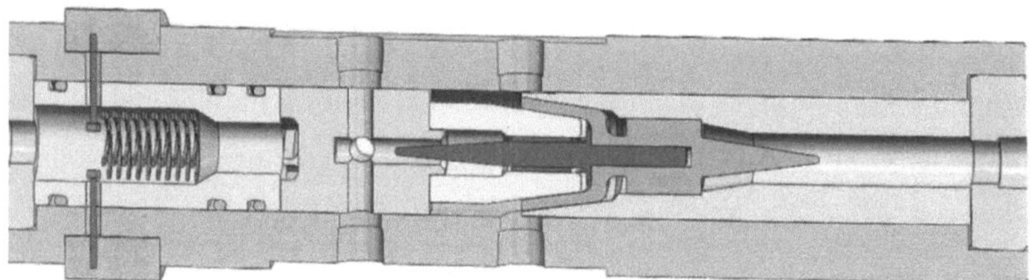

Figure 18. Fuel mixture activation device (in cross-section), in which a resonance sensor is installed in the output section, tuned to a certain system of parameters that allow monitoring the proportions between the components of the fuel mixture or fuel emulsion.

19. Periodic operation principle

- switch from continuous to periodic (pulsed) operation;

- if the action is already performed periodically, change the frequency;

- use the pauses between pulses for another action.

The fuel mixture activation device can equally function in any mode because it has no moving parts and is always ready for action (This also applies fully to the following two principles).

20. The principle of continuity of utility

• operate continuously (all parts of the facility must be operating at full load at all times);

• operate continuously (all parts of the facility must be operating at full load at all times);

21. The principle of slippage

• Conduct the process or individual steps (e.g. harmful or hazardous) at high speed.

22. The principle of "turning harm into good"

• utilise harmful factors (in particular environmental hazards) to produce positive effects;

• eliminate the harmful factor by compounding it with other harmful factors;

• amplify the harmful factor to the point where it is no longer harmful.

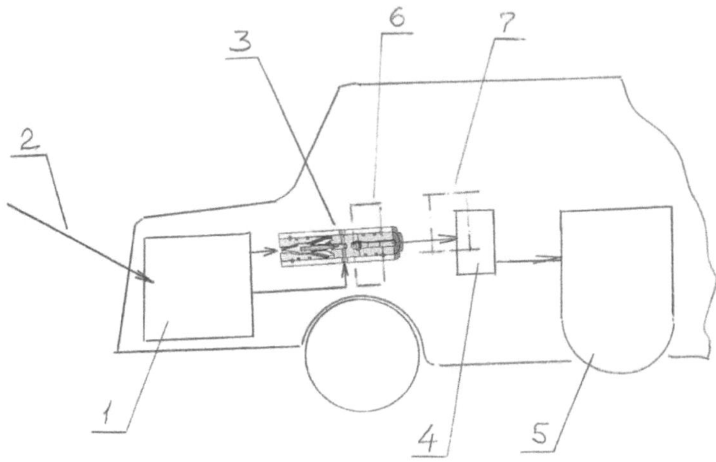

Figure 19. The process of fundamental change of vehicle layout, which allows to turn the harm of the need to increase the number of nodes and parts - in favour. And the transformation of the vehicle in full compliance with the principles - smart vehicle.

The numbers in the figure denote:

1 - fuel tank

2 - introduction of fuel composition into the fuel tank

3 - device for complex activation of fuel mixture

4 - high-pressure pump

5 - vehicle engine

23.   Feedback principle

- introduce feedback;
- if there is feedback, change it.

This principle is nowadays unambiguously realised in control and monitoring systems, including smart vehicles, in addition to the following three principles

24. The "middleman" principle

- Use an intermediate object that carries or transfers the action;

- temporarily attach another (easily removable) object to the object.

25. Self-service principle

- the facility should be self-sustaining with auxiliary and repair operations;

- utilise waste (energy, substances).

26. Copying principle

- Use simplified and cheaper copies of an inaccessible, complex, expensive, inconvenient or fragile object instead of an inaccessible, complex, expensive, inconvenient or fragile object;

- replace an object or a system of objects with their optical copies (images). Use in this case change of scale (increase or decrease copies);

- if visible optical copies are used, switch to infrared and ultraviolet copies

27. Cheap non-durability in lieu of expensive durability

- replace an expensive object with a set of cheap objects, while sacrificing some qualities (e.g. durability).

Long periods of pilot operation of fuel and fuel mixture activation devices have shown that the selected design option and the determining

the technical characteristics of the device ensure high durability at optimum cost

This, of course, depends on the technical characterisation factor that determines whether the entire production cycle of the device can be realised without the use of moving parts

28. Replacement of the mechanical system

* replace mechanical circuitry with optical, acoustic or "smell" circuitry;

* use electric, magnetic and electromagnetic fields to interact with an object;

* to move from fixed to moving fields, from fixed to time-varying fields, from unstructured to structured fields;

* use fields in combination with ferromagnetic particles.

Everyone knows that the most reliable objects are those that have no moving parts

Since one factor in the technical characterisation of a device for activating fuels and fuel mixtures is the absence of moving parts, it is possible to adopt the principles of design, manufacture and application of such devices corresponding to the general principle of no replacement of the mechanical system.

29. Use of pneumatic and hydraulic structures

* use gaseous and liquid parts instead of solid parts of the object;

* use electric, magnetic and electromagnetic fields to interact with an object: inflatable and hydro-inflatable, air cushion, hydrostatic and hydrojet.

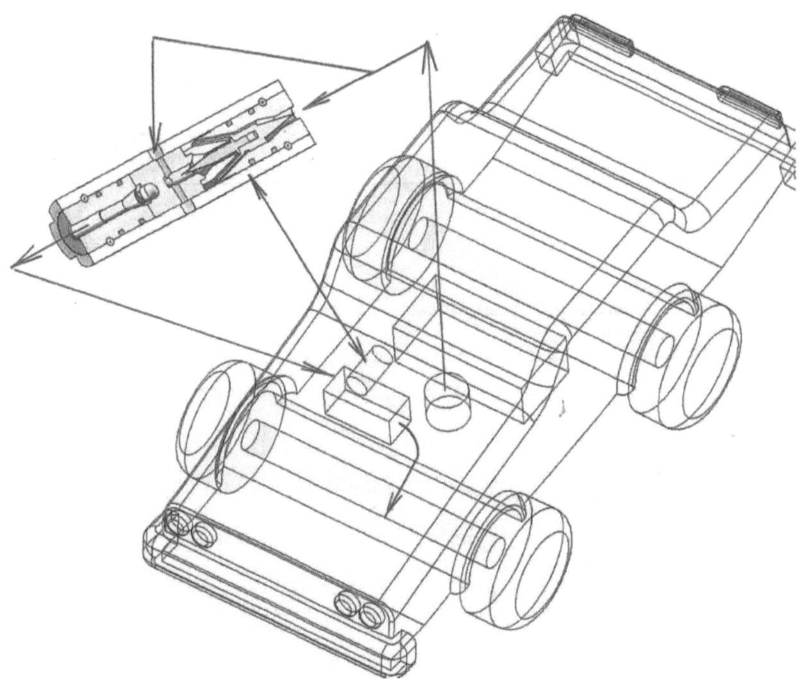

Figure 20. Transparent model of a smart vehicle, wherein, for the purposes of optimising the operating cycle and intensifying the combustion processes and ensuring that the combustion process obtains all the necessary properties and characteristics to include such a vehicle in a smart transport system, the fuel system includes hydraulic structures combining in its characteristic and pneumatic functions.

30.   Use of flexible shells and thin films

- use flexible shells and thin films instead of conventional structures;

• isolate the object from the external environment using flexible shells and thin films.

As an example of the use of the principle, thin-film microassemblies in microelectronics, which have significantly increased the speed of performance and, in turn, this increase in performance has allowed the transition to smart systems, including smart transport in general and smart transport systems in particular

31. Application of porous materials

• make the object porous or use additional porous elements (inserts, coatings, etc.);

• if the object is already porous, pre-fill the pores with some substance.

The emergence of composite materials, including those with pseudoporous structure, has allowed at the design stage to clearly predict and model the role of porosity or pseudoporosity in the formation of technical characteristics and output parameters of the designed device or product.

Very often the emergence of new products with unusual properties and characteristics is due to the use of porous or pseudoporous materials and components.

32. Principle of colour change

• change the colouring of an object or the external environment;

• change the degree of transparency of an object or the external environment.

Modern computer-aided design techniques allow for colour variations and changes in transparency during design or during computer simulation.

33. Principle of homogeneity

• objects interacting with this object must be made of the same material (or close to it in properties).

Homogenisation of fuel mixtures or other equivalent materials practically follow the homogeneity principle.

34. Principle of discarding and regeneration of parts

• The part of the object that has fulfilled its purpose or has become unnecessary must be discarded (dissolved, vaporised, etc.) or modified directly during the work;

• expendable parts of the facility must be recovered directly from the work. The methods and systems of aerodynamic and hydrodynamic mixing, dynamic emulsion preparation and online homogenisation are not characteristic of this principle, as they are based on the principle of not having any regeneration and waste part rejection processes.

35. Change of physical and chemical parameters of the object

• change the aggregate state of the object;
• change the concentration or consistency;
• change the degree of flexibility;
• change the temperature.

This principle is used in various variations in the preparation of fuel mixtures and fuel emulsions, as well as in homogenisation processes. This principle can be considered wider, if we take into account the fact that the formation of fuel capsules creates new properties for the fuel mixture, and the formation of fuel mixture with simultaneous foaming with the help of foam generator creates the effect of compressible liquid instead of incompressible liquid. This principle is directly related to the following principles - application of phase transitions and application of thermal expansion.

36. Application of phase transitions

• Use phenomena arising from phase transitions, e.g. volume change, release or absorption of heat, etc.

37. Application of thermal expansion

• utilise the thermal expansion (or contraction) of materials;

- use several materials with different coefficients of thermal expansion.

38. Use of strong oxidising agents

- to replace the normal air with enriched air;

- to replace the enriched air with oxygen;

- use ozonised oxygen;

- replace ozonated oxygen (or ionised oxygen) with ozone.

This principle is only applicable in chemical and electrochemical treatment processes, but in transport systems where the fuel is activated in foam generators, this principle can be applied.

39. Application of inert medium

- to replace the normal medium with an inert medium;

- to conduct the process in a vacuum.

This principle in today's innovation practice is implemented in the framework of typical computer modelling, and with the right questions and detailed technical specifications for the relevant models of the innovation system, this process can be carried out much more accurately and it is quite realistic to obtain information allowing to discover or anticipate new options and new directions with a higher level of novelty and efficiency.

40.   Application of composite materials

- move from homogeneous materials to composite materials.

It is well known that the substitution of one material for another is not recognised as an invention. The composite material itself is the subject or object of an original invention, but very often, when an ordinary construction material is changed to a composite, the properties and possibilities of the product are so changed that the product in which composites are used becomes completely new, unknown before, with completely new functions and technical characteristics.

Of course, in order to make such a replacement, it is necessary to carry out such a volume of work, which is comparable to the development of a fundamentally new technology or a fundamentally new product, and this is only possible for companies with powerful research departments. In smart transport systems not only composite structural materials, but also consumables: fuel, for example, in the form of an emulsion or a mixture. A homogenised fuel with an equivalent viscosity level both at the centre of the flow in the pipeline and at the periphery of the flow in the pipeline is also a material with composite properties. Thus, the introduction of an online homogenisation device, or an online emulsion preparation device, or an online fuel blend preparation device into the fuel system is equivalent to the use of composite materials and has a corresponding effect in smart transport.

In the process of analysing the situation, let us return to the **tools of TRIZ and ARIZ,** which were created to overcome this kind and complexity of contradiction complexes.

In the theory of inventive problem solving there is a special programme for solving difficult problems. This programme breaks down the solution process into about 50 consecutive steps. The programme is equipped with special steps that help to overcome psychological inertia. The programme also has a rich information support. The programme is called ARIZ, an algorithm for solving inventive problems.

Initially, the "methodology of invention" was conceived as a set of rules such as: "to solve a problem means to find and overcome a technical contradiction" or "the solution of a problem is the stronger the lower the expenditure of matter, energy, space, and time". The emerging "methodology of invention" also included some typical techniques: fragmentation, unification, inversion, change of aggregate state, replacement of mechanical scheme with chemical one, etc.

The main source for identifying rules and techniques was information on the work of great inventors, their own inventive practice, and materials on the history of technology.

By the mid-50s, a belief had developed and gained ground that inventors, even the strongest ones, work by the inefficient method of trial and error, and therefore the endeavour to discover and exploit the "secrets of creativity" is futile.

It is necessary to build a fundamentally new "methodology of invention" based on the use of objective laws of technical systems development. These laws can be revealed by systematic analysis of large arrays of patent information.

By the end of the 1950s, it became clear that the "methodology of invention" should include not only ARIZ, but also a section on the laws of development of technical systems and a constantly growing information fund. The "methodology of invention" was to give way to the "science of invention". This idea met with strong resistance. The "methodology of invention" was looked upon as something more or less tolerable: after all, it was useful recommendations based on the study of inventors' experience, there was no open subversion of "holy" concepts. "The Science of Invention" swung at the "sacred" - denying the exceptionalism of great inventors, touching on the familiar notion of the incognisability of the creative process. "Methodology of Invention" helped to "illuminate" - the "science of invention" denied all old technology, denied natural ability. This was at that time pure heresy.....

Over the years, the Programme has become more rigid and defined. In the process of analysis, the operational area and the conflicting demands placed on it (a prototype of the OP) are defined. The RWS operator has been introduced. The table of TP elimination is finalised, the list of techniques is enlarged (first 40, then 50). Prescriptions for steps, notes, examples are introduced. The main operators form a system - the interrelation between the steps has been strengthened, there is a new part - preliminary evaluation of the found idea.

But also over the years the situation has changed fundamentally, as processor technology has appeared, which has made us treat the concept of reliability in a completely different way, as the rigid, mechanical understanding of reliability has been replaced by a more flexible one, due to a more precise process and work cycle control using the analytical and control capabilities of processor technology.

The following main **directions in the evolutionary development of TRIZ and ARIZ** in terms of synthesis and modification of complex technical solutions, one of the fundamental basic indicators of which is the integrative reliability of the system:

1.   Traditional for the evolution of ARIZ is a general increase in the degree of algorithmicisation due to a more complete and deeper use of objective laws of development of technical systems, including processor systems.

2.   Significant strengthening of the "bridge" between the physical contradiction and the way of its resolution, including that based on the use of composite materials and the latest achievements of digital technology.

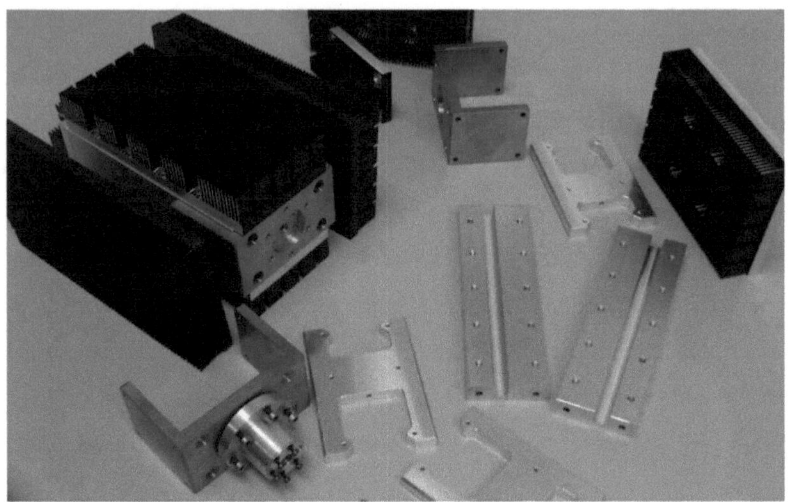

Figure 21. Innovative device for mounting a high-power laser diode.

The device design uses composite materials based on spherical capsules with two levels, where a copper-coated artificial diamond core is located in the centre.

This structure dissipates heat and cools the laser diode extremely efficiently. This is a clear example of how the introduction of a composite material into the design allows the application of high-energy laser diodes in, for example, medical technology without the creation of special cooling systems and a significant increase in the cost of the process.

3. Strengthening the information foundation, strengthening the links between ARIZ and standards, including the combination of operational production standards with environmental standards, the requirements and limitations of which go against traditional economic norms.

4. Preferential development of the second half of ARIZ (development and use of the found idea) into an independent algorithm with components of the type: Programme, system and method or Apparatus, device, system, programme and associated method

5. Development of a new initial part (or a separate algorithm) to identify

new compositional and integrative problems.

6. Strengthening the general educational function. ARIZ should more vigorously develop the skills of strong, complex thinking.

7. Gradual increase in versatility in the process of creating a compositional model of an apparatus or process closely related to software and processor technology.

List of used literature and patent materials

Annex 1

United States Patent Application

20190069225

Kind Code

A1

NATHANSON; Martin

February 28, 2019

VEHICLE COMMUNICATIONS VIA WIRELESS ACCESS VEHICULAR ENVIRONMENT

Abstract

A vehicle heads-up display (HUD) includes a *smart* phone or tablet computer with at least one processor running at least one computer programme adapted to enable the HUD to: (i) establish a connection to an on board unit (OBU) of the vehicle using the mechanisms provided by ICMPv6 for IPv6 router discovery, and acquire an IPv6 address through the mechanism of Stateless address auto configuration (SLAAC); (ii) process an authentication challenge from a Roadway Authorisation Server (RAS); and (iii) respond to an authentication challenge from said Roadway Authorization Server (RAS).

Annex 2

United States Patent Application

Kind Code

Malkes; William A. ; et al.

20190051167

A1

February 14, 2019

SYSTEM AND METHOD OF ADAPTIVE TRAFFIC MANAGEMENT AT AN INTERSECTION

Abstract

A traffic control system and a method of automatic zone creation and modification for a *smart* traffic camera to be used in adaptive traffic management at an intersection are disclosed. One aspect of the present disclosure is a method including applying default zone parameters to define detection zones at one or more sensors installed at an intersection, the detection zones being used by the one or more sensors for monitoring and detecting traffic conditions at the intersection; determining a current vehicular traffic flow rate and a current pedestrian traffic flow rate at the intersection; determining if a triggering condition for adjusting one or more of the default zone parameters; and adjusting the one or more of the default zone parameters if the triggering condition is met.

Annex 3

United States Patent Application

Kind Code

Corpus; Roy C. ; et al.

20190027018

A1

January 24, 2019

ARTIFICIAL INTELLIGENCE BASED SERVICE CONTROL AND HOME MONITORING

Abstract

In some examples, artificial intelligence-based service control and home monitoring may include ascertaining, from a monitoring tool, an alert related to operation of a device or service monitored by the monitoring tool, and generating, based on the alert, a support call that includes a phone call, an e-mail, a Short Message Service (SMS), and/or a *smart* device notification to a support personnel. Based on an issue addressed in the alert, an incident ticket may be generated, and based on a response to the support call, and determination of a resolution to the issue addressed in the alert, the incident ticket may be modified to include the resolution. Further, a service level agreement may be analysed, and based on an analysis of the alert, the support call, the incident ticket, and the service level agreement, metrics related to the resolution to the issue addressed in the alert may be generated.

United States Patent Application

20190012909

Kind Code

A1

MINTZ; Yosef

January 10, 2019

SYSTEM AND METHODS TO APPLY ROBUST PREDICTIVE TRAFFIC LOAD BALANCING CONTROL AND ROBUST COOPERATIVE SAFE DRIVING FOR SMART CITIES

Abstract

Apparatuses, systems and methods applying an innovative non-discriminating and anonymous car related navigation driven traffic model predictive control, producing predictive load-balancing on road networks which dynamically assigns efficient sets of routes to car related navigation aids and which navigation aids may refer to in dash navigation or to *smart* phone navigation application. The system and methods are may enable, for example, to improve or to substitute commercial navigation service solutions, applying under such upgrade or substitution a new highly efficient proactive traffic control for city size or metropolitan size traffic.

United States Patent Application

20180376305

Kind Code

A1

Ramalho de Oliveira; Patricia Cristina

December 27, 2018

METHODS AND SYSTEMS FOR DETECTING ANOMALIES AND FORECASTING OPTIMISATIONS TO IMPROVE SMART CITY OR REGION INFRASTRUCTURE MANAGEMENT USING NETWORKS OF AUTONOMOUS VEHICLES

Abstract

Methods and systems are provided for detecting anomalies and forecasting optimisations to improve *smart* city or region infrastructure management using networks of autonomous vehicles. An autonomous vehicle may receive initial information relating to infrastructure utilised by a plurality of autonomous vehicles, and may acquire, during operation in the infrastructure, real-time information relating to the infrastructure and/or to other ones of the plurality of autonomous vehicles. The acquired information may be processed, and based on the processing of the acquired information and the initial information, anomalies and/or problems affecting the infrastructure and/or operation of the plurality of autonomous vehicles in the infrastructure may be detected.

Annex 6

United States Patent Application

20180330307

Kind Code

A1

Anderson; Evelyn R. ; et al.

November 15, 2018

METHOD AND SYSTEM TO ASSIST PEOPLE TRAVELLING TOGETHER THROUGH A TRANSPORT HUB

Abstract

The present invention is a method and system to assist people travelling together through a transport hub from a current location to a destination location using a *smart* portable personal computing device which provides personal data, travel data and transport hub data. The method provides for sub-groups of travellers having common travel preferences, and it maps routes according to those preferences for the group members to travel through the transport hub together. Continuous monitoring of values provides an alert on each sub-group member's personal computing device when the personal data, travel data, transport hub data, or predetermined parameters change.

United States Patent Application

20180330294

Kind Code

A1

Anderson; Evelyn R. ; et al.

November 15, 2018

PERSONAL TRAVEL ASSISTANCE SYSTEM AND METHOD FOR TRAVELLING THROUGH A TRANSPORT HUB

Abstract

The present invention is a method and system to assist a person travelling through a transport hub from a current location to a destination location with a *smart* portable computing device which provides personal data, travel data and transport hub data. The method maps a route according to personal preferences for the person to travel through the transport hub. Continuous monitoring of values provides an alert when the personal data, travel data, transport hub data, or predetermined parameters change.

Annex 8

United States Patent Application

20180328079

Kind Code

A1

LIM; Chee Kean ; et al.

November 15, 2018

SMART SECURITY DEVICE AND SYSTEM

Abstract

A *smart* security device comprising: a housing, and electronics provided in the housing, the electronics comprising a tag uniquely associated with the *smart* security device and configured to automatically and periodically emit a first signal, the first signal being a Bluetooth signal configured to be sent and received via a Bluetooth low energy wireless personal area network; and a NB-IoT module configured to send and receive signals via a NB-IoT network.

Annex 9

United States Patent Application

20180308045

Kind Code

A1

ARENA; DAVID

October 25, 2018

MOBILE APPLICATION WITH ENHANCED USER INTERFACE FOR EFFICIENTLY MANAGING AND ASSURING THE SAFETY, QUALITY AND SECURITY OF GOODS STORED WITHIN A TRUCK, TRACTOR OR TRAILER AND ASSESSING USER COMPLIANCE WITH REGULATIONS AND QUALITY OF PERFORMANCE

Abstract

A system and method for ensuring the safety of goods transported via highway, particularly humanly consumable goods, is taught by the present invention. Three main aspects include a *smart* phone hub, a portable sensor for monitoring the transported goods and a physical locking mechanism to lock the trailer. According to the present invention, a truck driver uses a *smart* phone to interface between a trailer payload supervisor and the payload itself, to insure the safety of the transported goods. Upon loading a trailer, a truck driver uses a smartphone to activate an internal sensor and snap an image of the locked trailer (with an electronic padlock and a licence plate, for example). According to the present invention, an enhanced user profile includes information or ratings based on past driver performance and a loyalty programme to reward driver's effective and safe (perishable cargo safety) usage of the invention.

United States Patent Application

Kind Code

Evans; Michael Steward

20180320402

A1

November 8, 2018

Intelligent POD Management and Transport

Abstract

An exchange station has openings for drones, a passenger check-in/check-out bay for processing passengers, a drone connect/release bay, having apparatus adapted to manage passenger pods mounted on *smart chassis*, and a computerised control system in wireless communication with control circuitry in the drones and *smart* chassis, guiding *smart chassis* with mounted passenger pods and drones, to make the exchange of pods from the *smart chassis* to drones. A passenger entering the passenger check-in bay is loaded into a pod mounted on a *smart chassis*, the pod with passenger is transported to the drone-connect/release bay, and the pod is there joined to a bare drone and disconnected from the *smart chassis*, the drone leaving with the passenger pod to a destination, and the *smart chassis* travelling away from the drone connect-release bay.

Annex 11

United States Patent Application

Kind Code

MAIMON; MOTI

20180308087

A1

October 25, 2018

SYSTEM AND METHOD FOR MANAGEMENT OF A SMART OBJECT

Abstract

A system and method for performing an operation on a *smart* object, the system including at least one user interface, having access to the Internet, including a *smart* object reader/writer and a processor, a universal, multi-platform "dumb" SDK (Software Development Kit), supporting multi-platform communication, embedded in each user interface, a central management unit disposed in the Cloud, configured to operate under multiple platforms and to communicate over multiple standards, in two-way communication with each *smart* object via its SDK, at least one provider backend in communication with the at least one user interface and with the central management unit, a secure encryption unit proving a physical key for each communication between each SDK and the central management unit, and a web services managing server configured to provide secure two-way communication between each user interface, the secure encryption unit, the provider backend and the central management unit.

United States Patent Application

20180315011

Kind Code

A1

Clarke; John ; et al.

November 1, 2018

Limited Spatial Digital Directory with Physical Navigation for Optimizing Smart Carts **Abstract**

The present disclosure relates to a programme or system for navigating a store or handling and locating inventory within a store. This application has several embodiments including but not limited to those where the programme operates on a computerised console directly docked on a store cart, a mobile device tethered to a store cart, or any combination of the two. Security features, charging features (using energy including but not limited synergistically charged batteries or solar panels) scanning features, geospatial features and other peripheral device enabled features are disclosed in the multiple embodiments of the disclosure. For use with a *smart* cart the invention disclosed herein can greatly enhance a shopping experience, allowing for efficient shopping, with features like phone charging, simplified item location and purchasing for shoppers and ease of inventory management, cart-maintenance, and security for shop keepers.

Annex 13

United States Patent Application

Kind Code

Tran ; Bao ; et al.

20180326291

A1

November 15, 2018

SMART DEVICE

Abstract

An Internet of Thing (loT) device includes a transceiver coupled to a processor.

Blockchain *smart* contracts can be used with the device to facilitate secure operation.

yes
I want morebooks!

Buy your books fast and straightforward online - at one of world's fastest growing online book stores! Environmentally sound due to Print-on-Demand technologies.

Buy your books online at
www.morebooks.shop

Kaufen Sie Ihre Bücher schnell und unkompliziert online – auf einer der am schnellsten wachsenden Buchhandelsplattformen weltweit! Dank Print-On-Demand umwelt- und ressourcenschonend produzi ert.

Bücher schneller online kaufen
www.morebooks.shop

info@omniscriptum.com
www.omniscriptum.com

Printed by Books on Demand GmbH, Norderstedt / Germany